PHYSICAL SCIENCE

ice to steam

Changing States of Matter

Penny Johnson

Rourke
Educational Media
rourkeeducationalmedia.com

www.rourkeeducationalmedia.com

PHOTO CREDITS: p. 39: Douglas Allen/istockphoto.com; p. 19:Kelly Cline/istockphoto.
com; pp. 29, 42: Corbis; p. 8: Liz Friis-Larsen/istockphoto.com; p. 6 right & 7: Lise Gagne/
istockphoto.com; p. 13: Amn Matthew J. Hannen/USAF/Department of Defense; p.
15: David Hernandez/istockphoto.com; pp. 6 middle, 16, 18: istockphoto.com; p. 27:
Steve Klaver/Star Ledger/Corbis; p. 33: Jacques Langevin/CORBIS SYGMA; p. 41: David
Lewis/istockphoto.com; p.43: Xavier Marchant/istockphoto.com; p. 26: Jamie Marland/
istockphoto.com; p. 24: James Martin/Getty Images; p. 4: NASA; p. 12: NOAA;
p. 6 left Christopher O'Driscoll/istockphoto.com; p. 20: Rey Rojo/istockphoto.com;
p. 25: Cedric H. Rudisill/USAF/Department of Defense; p. 40: Clive Sanders/EASI-Images/
CFWImages.com; p. 23: Kit Sen Chin/istockphoto.com; p. 17: Jordan Shaw/istockphoto.com;
p. 21: Matt Tilghman/istockphoto.com; p. 31: R.I. Tiling/U.S. Geographical Survey;
p. 5: Tony Tremblay/istockphoto.com; p. 11: Michael Valdez/istockphoto.com; p. 22: Graça
Victoria/istockphoto.com; p. 32: Roger Werth/Time & Life/Getty Images; p. 30: Shannon
Workman/istockphoto.com.

Cover shows a snowy landscape in the Austrian Alps [Ingmar Wesemann/istockphoto.com].

Produced for Rourke Publishing by Discovery Books
Editors: Geoff Barker, Amy Bauman, Rebecca Hunter
Designer: Ian Winton
Cover designer: Keith Williams
Illustrator: Stefan Chabluk
Photo researcher: Rachel Tisdale

Library of Congress Cataloging-in-Publication Data

Johnson, Penny.
  Ice to steam : changing states of matter / Penny Johnson.
     p. cm. -- (Let's explore science)
  Includes index.
  ISBN 978-1-60044-605-4 (hard cover)
  ISBN 978-1-61236-231-1 (soft cover)
  1. Matter--Properties--Juvenile literature. I. Title.
  QC173.36.J64 2008
  530.4--dc22
                            2007020111

Printed in China, FOFO I - Production Company
  Shenzhen, Guangdong Province

rourkeeducationalmedia.com

customerservice@rourkeeducationalmedia.com • PO Box 643328 Vero Beach, Florida 32964

Contents

chapter one

Ice to Steam

Look at this picture of Earth. You can see the continent of Africa. The rest of what you see is water in the form of oceans, **clouds**, and ice.

Solids, Liquids, and Gases

The oceans are made of **liquid** water mixed with salt and other chemicals. The clouds you can see are tiny drops of liquid water floating in the air.

There is ice around the poles because it is very cold there. Ice is **solid** water. It is the same material as liquid water, but it has different properties because it is a solid.

The air around Earth is a **gas**. It contains **water vapor**. This is water that has become a gas. We cannot see water vapor.

We can describe all the things around us as solids, liquids, or gases. Each kind of material has different properties.

Solids have fixed shapes. They have fixed **volumes**, too. You would need force to try to change solids. Liquids do not have fixed shapes. You can pour them. Liquids have fixed volumes. Gases do not have fixed shapes or volumes. Gases spread out to fill the space they are in.

▲Ice is a solid material.

Changing State

Ice, liquid water, and water vapor are three different **states** of water. You can change water from one state to another by changing the temperature.

If you heat ice, it turns into liquid water. This is called **melting**. Melting happens when the temperature is 32°F (0°C) or higher. This is the **melting point** of water.

If you cool liquid water to 32°F (0°C) or lower, it becomes solid. This is called **freezing**. The melting and **freezing points** of a material are the same temperature.

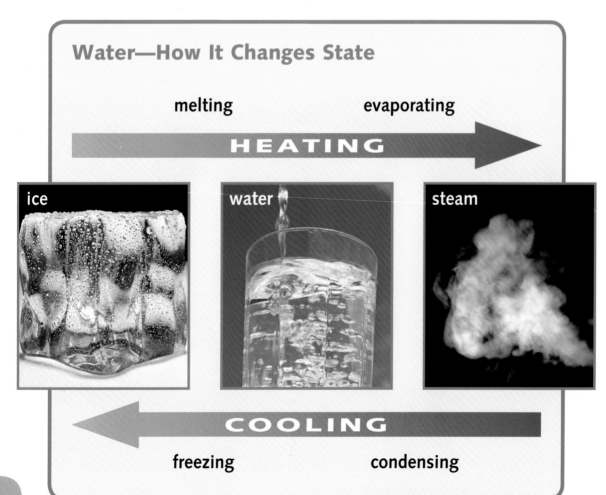

Water—How It Changes State

melting   evaporating

HEATING →

ice   water   steam

← **COOLING**

freezing   condensing

INVISIBLE STEAM

Steam is an invisible gas. You cannot see steam. What you see when a kettle boils is actually tiny drops of water.

Liquid water can turn into a gas at any temperature. We say it **evaporates**. Evaporation happens fastest when water is heated to 212°F (100°C). This is the **boiling point** of water. When water is **boiling**, we call the hot water vapor **steam**.

When steam or water vapor cools down, it turns back to liquid water. This is called **condensing**.

chapter two

Water and the Weather

Some of the water you drink comes from rivers. Some comes from rainwater that has soaked into the ground. All of this water originally came from the ocean. Eventually, it will go back to the ocean. This movement of water is called the **water cycle**.

WHO HAS ALREADY DRUNK YOUR WATER?

You may have drunk some of the same water that Abraham Lincoln drank! This is because water is always moving around the water cycle.

The Water Cycle

4. If it is cold, the water in the clouds freezes and forms snow.

3. Water droplets join up and get bigger. They fall as rain.

2. The water vapor cools and condenses. The droplets of water form clouds.

1. Heat from the Sun makes water evaporate from the ocean.

5. Rain runs into rivers and streams. Water in rivers runs into the ocean. Water is stored in lakes and reservoirs.

6. Some water soaks into the ground. We can dig wells to get this water.

How Does Water Get to Your Home?

The water that you use every day has come from a river or from a well. It is brought to your home through many miles of pipes.

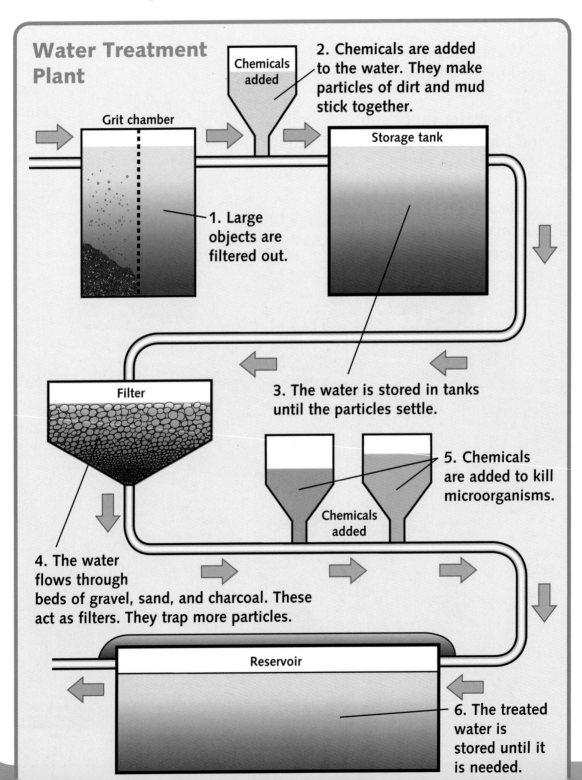

Water Treatment Plant

Chemicals added

2. Chemicals are added to the water. They make particles of dirt and mud stick together.

Grit chamber

Storage tank

1. Large objects are filtered out.

Filter

3. The water is stored in tanks until the particles settle.

5. Chemicals are added to kill microorganisms.

Chemicals added

4. The water flows through beds of gravel, sand, and charcoal. These act as filters. They trap more particles.

Reservoir

6. The treated water is stored until it is needed.

DRINKING SEAWATER

Most of the water on Earth is in the oceans. This water is salty, so we cannot drink it. Some dry countries have no rivers. Their drinking water comes from the ocean. Water from the ocean must be treated in a desalination plant to remove the salt. This costs a lot more money than cleaning up water from rivers.

Water from a river may be very muddy. It may contain **microorganisms** that could make you ill. The water has to be treated to make sure it is safe for you to drink.

Water from sinks, showers, and toilets is taken away in sewage pipes. This is called wastewater. Wastewater must be treated before it runs back into rivers or the sea. This stops the waste from causing **pollution**.

▼Wastewater is made safe at a water treatment plant.

Extreme Weather

Clouds form when water vapor in the air condenses. Sometimes clouds grow into huge cumulonimbus clouds. These clouds can bring bad weather.

There are strong winds blowing up and down inside cumulonimbus clouds. It is dangerous for airplanes to fly through them. The winds inside the cloud also cause thunder and lightning.

LIGHTNING

In the United States, lightning kills more than 70 people each year.

▲Lightning from a cumulonimbus cloud.

▲Cumulonimbus clouds can bring bad weather. This photograph shows the results of severe flooding in Grand Forks, North Dakota.

Cumulonimbus clouds are very big. They hold a lot of water. Rain from these clouds can cause floods. If part of the cloud temperature is below the freezing point of water, hail stones can form.

chapter three

Solids, Liquids, and Gases at Home

All the materials around you can be described as solids, liquids, or gases.

Melting Points

Different materials melt at different temperatures. The chart shows the melting points of some materials you know. You might find some of these materials in your kitchen.

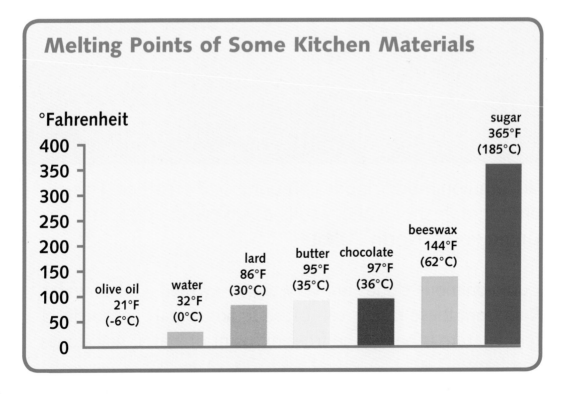

Melting Points of Some Kitchen Materials

°Fahrenheit

sugar
365°F
(185°C)

beeswax
144°F
(62°C)

lard
86°F
(30°C)

butter
95°F
(35°C)

chocolate
97°F
(36°C)

olive oil
21°F
(-6°C)

water
32°F
(0°C)

400
350
300
250
200
150
100
50
0

▶Chocolate melts at about 97°F (36°C), which is just below body temperature. Chocolate feels good to eat because it melts in your mouth!

The temperature in a room is usually around 65°F (18°C). A material with a melting point lower than this will be a liquid. Olive oil and water are liquids at room temperature.

Boiling Points

Different materials also boil at different temperatures. Liquids such as milk and fruit juice are mostly water. This means they should boil at about the same temperature as water does. So they boil at around 212°F (100°C).

Boiling Points of Some Kitchen Materials

°Fahrenheit

- 600
- 500
- 400
- 300
- 200
- 100
- 0

nail polish remover 133°F (56°C)

alcohol 172°F (78°C)

water 212°F (100°C)

olive oil 572°F (300°C)

COOKING

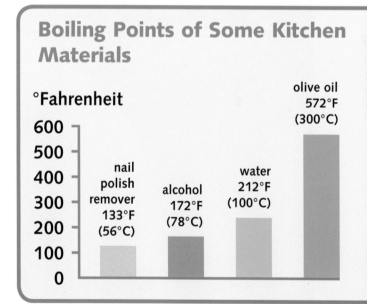

Heating butter makes it melt. It is turned into a liquid. Heating an egg makes it turn from a liquid to a solid. This happens because the heat is changing the chemicals that the egg is made from. It is not the same as a change of state.

Is Sugar a Solid or a Liquid?

Solids have fixed shapes. Liquids flow and take the shape of the container you put them into. The surface of a liquid is flat. Sugar has some of the **properties** of a liquid. You can pour it from one container into another. It will take the shape of the container you put it in.

▲Sugar has some of the properties of a liquid. It also has some of the properties of a solid.

SPRAYING POWDER

Some fertilizers come in a powder form. These can be pumped and sprayed, just like a liquid.

Sugar also has some of the properties of a solid. You can make a pile of sugar. But it does not end up with a flat surface unless you smooth it out.

Look closely at the sugar. You can see that it is made of tiny **crystals**. The crystals are solid. Solid materials sometimes act like liquids if the pieces are very small. Other solid materials that can act like liquids are salt and sand.

Mixtures of States

Many things around you are combinations of materials in different states.

Some hairspray comes in a spray can. Hairspray is a mixture of tiny drops of liquid and a gas. The liquid is a chemical that keeps your hair in place. The gas helps spray the drops of liquid over your hair. A mixture with drops of liquid in a gas is called an **aerosol**.

CLOUDS

Clouds are aerosols because they are tiny drops of liquid (water) mixed with a gas (the air). Fogs and mists are also aerosols. They are just clouds that reach the ground!

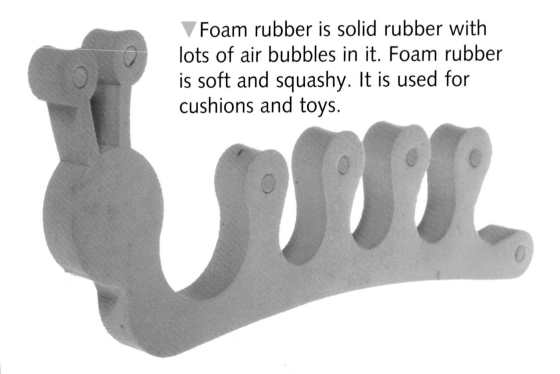

▼Foam rubber is solid rubber with lots of air bubbles in it. Foam rubber is soft and squashy. It is used for cushions and toys.

▲ A foam (cream) on top of an emulsion (milk).

Some foods are also mixtures. Milk is a mixture of water and droplets of liquid fat. Mixtures of two liquids are called emulsions. Whipped cream is a foam of air bubbles in a liquid.

chapter four

Evaporation

How Can You Speed Up Evaporation?

Your clothes need to be washed when they get dirty. After they are washed, the water in them evaporates. When all the water has evaporated, the clothes are dry.

You can help wet clothes dry quickly by spreading them out on a clothesline. Water evaporates faster when it is warmer. So the clothes will dry faster on a warm day than on a cold day.

▼These aprons would dry even faster if there was a breeze.

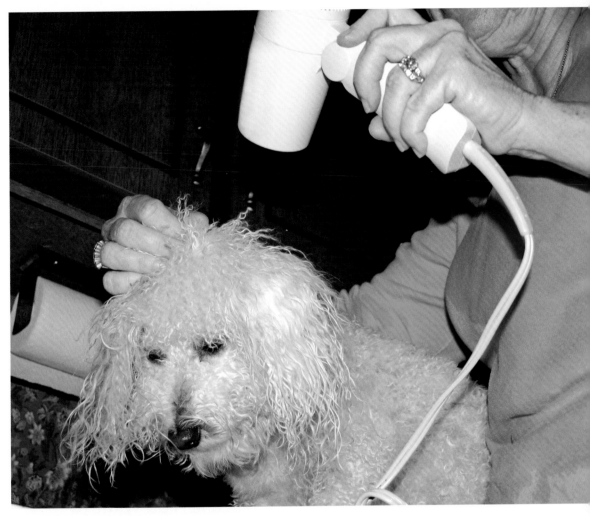

▲The dog will dry quickly because of the warm air from the hair dryer.

When water evaporates, the water vapor mixes with the air. Eventually the air has as much water vapor as it can hold. When this happens, no more water can evaporate from the clothes.

Wet clothes dry faster on a windy day. This is because the wind blows the water vapor away. Clothes dryers and hair dryers also speed up evaporation. They blow warm air. The warmth helps evaporate water from the clothes or hair. The moving air carries the water vapor away.

▲You lose a lot of water as sweat in hot weather. You need to drink more water to replace it. You want to make sure your body has enough.

How Does Evaporation Keep You Cool?

When you run around, your body gets warm. You may start to sweat. Sweat is your body's way of keeping cool. Sweat glands in your skin produce this salty liquid. As the sweat evaporates, it cools your skin. When some of a liquid evaporates, the liquid left behind is cooler than it was to start with.

Sweat works best on a breezy day. Liquids can evaporate faster when a breeze moves away the vapor that has already evaporated.

Humid air is air that already contains a lot of water vapor. This makes it harder for sweat to evaporate. Humid days feel uncomfortable because your sweat does not evaporate. You feel hot and sticky.

COOL DOGS!

Dogs can sweat only through their feet, where there is no fur. Dogs keep cool by panting.

chapter five

Extreme Changes of State

Metals such as iron and steel can be made into different shapes. Some iron shapes are made by melting the iron. The liquid iron is then poured into a mold. When it cools down, it becomes a solid again. The solid takes the shape of the mold.

▼ Molten iron being poured into a mold.

Some objects get very hot when they are used, so they must be made of materials with high melting points. That way, they do not melt or change shape while they are being used. The melting point of the material must be higher than the temperature the object will reach.

The melting points of some metals	
Metal	**Melting point**
aluminum	1,220°F (660°C)
gold	1,947°F (1,064°C)
iron	2,795°F (1,535°C)
silver	1,764°F (962°C)
steel	2,500°F (1,371°C)
titanium	3,020°F (1,660°C)

▲The temperature of a jet engine can get as hot as 2,800°F (1,538°C). The engine must be made of materials with high melting points. One such material is titanium.

Low Boiling Points

The air around you is a mixture of gases. These gases include **nitrogen**, **oxygen**, and **carbon dioxide**. All of these substances have very low boiling points. This means that even in the coldest weather, the temperature does not get low enough to make them condense into liquids.

Boiling Points of Some Gases

Gas	Boiling point
carbon dioxide	-108°F (-78°C)
methane	-263°F (-164°C)
oxygen	-297°F (-183°C)
nitrogen	-321°F (-196°C)

FREEZE-DRIED BODIES

In Sweden, a dead body can be frozen using liquid nitrogen. The body becomes very brittle. It is then vibrated, which breaks it up into a powder. The powder is buried. It forms compost that can be used by plants in less than a year.

◀Scientists use liquid nitrogen to freeze cells for use in research.

DRY ICE

Carbon dioxide is an unusual chemical. If you heat solid carbon dioxide, it does not become
a liquid. It changes straight into a gas. This change is called subliming. Solid carbon dioxide is known as "dry ice." It is used to make "smoke" effects on stage.

Some of the gases are liquids in other places in the solar system. Titan is one of the moons of Saturn. The temperature on its surface is about -289°F (-178°C). This is low enough for the gas methane to become a liquid. Pools of liquid methane are seen on the moon's surface.

chapter six

Volcanoes and Lava

The inside of Earth is very hot. In some places, it is hot enough to melt rock. Molten rock inside Earth is called **magma**. **Lava** is molten rock that comes out of volcanoes.

When the lava cools down, it forms solid rock. The lava that flows from Hawaiian volcanoes is very runny. It flows a long way before it cools. Gradually the new rock builds up a cone.

The lava from some volcanoes is more sticky. It does not flow very far before it cools. Sometimes, the volcano erupts by shooting bits of rock and ash into the air. The cooled lava and the ash build up steeper volcanoes.

EXTRATERRESTRIAL VOLCANOES

Volcanoes are found in other places in the solar system, too. Mars has some volcanoes that are inactive. They are said to be extinct. Lo is one of Jupiter's moons. It has active volcanoes.

▲Mount Fuji is a volcano in Japan. It is made of layers of lava and ash.

▲ Giant's Causeway in Northern Ireland is formed from about 40,000 columns of basalt. These volcanic rocks are made from cooled lava.

What Happens to Lava When It Freezes?

When molten rock, or lava, freezes, it forms crystals. If the rock freezes quickly, the crystals do not have time to grow very big.

Lava that runs out of volcanoes cools down quickly. It forms rocks like **basalt**. Basalt has tiny crystals.

Magma can get trapped underground. It takes a long time for magma to cool down, so the crystals have time to grow bigger. **Granite** is a rock that is formed when magma cools down underground.

▼This is a lava tube. Lava has frozen on the surface, but is still running along below.

▲A huge tower of smoke erupting from Mount St. Helen's in Skamania County, Washington.

Danger, Volcano!

Some volcanoes have deadly eruptions. One of the most famous is the eruption of Mount Vesuvius, in Italy. This happened in the year A.D. 79. It buried the towns of Pompeii and Herculanium in 50 feet (15 meters) of ash. More than 2,000 people were killed.

The biggest eruption in recent years was Tambora, in Indonesia. Tambora erupted in 1815. About 10,000 people died. It threw so much ash into the atmosphere that the whole Earth became cooler. That year was called "the year without a summer." The weather was so cool that the crops were lost. Another 20,000 people starved to death.

Lava and ash are not the only dangers from eruptions. The heat of the lava can melt snow and ice at the top of the volcano. The water forms mudflows. These can bury whole towns. Navado Del Ruiz is a volcano in Colombia. It erupted in 1985. The mudflows it caused killed more than 23,000 people.

▼ A mudflow in Armero, Colombia, caused by the eruption of Nevado del Ruiz in November 1985.

chapter seven

Particle Theory

Properties of Solids, Liquids, and Gases

All materials are made of tiny particles. These are called **atoms**. Sometimes atoms join together to form **molecules**. Scientists talk about different materials by talking about these particles. They talk about how the atoms (or molecules) are arranged. This helps scientists explain solids, liquids, and gases.

Solids

In solids, the particles are arranged in a regular pattern. They are held together by strong forces. This is why solids have a fixed shape. You cannot squash or change the volume of solids. The particles are already as close to each other as they can get.

▶This diagram shows the particles in a solid.

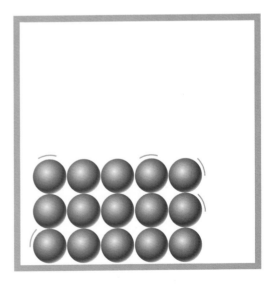

Liquids

With liquids, the particles are still very close. But the forces holding them together are not as strong as in solids. The particles can move around. This is why you can pour a liquid. You cannot squash a liquid because the particles are still very close together.

►This diagram shows the particles in a liquid.

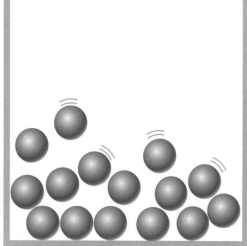

Gases

The particles in a gas are a long way apart. They move around very quickly. Gases are easy to squash because there is a lot of space between the particles.

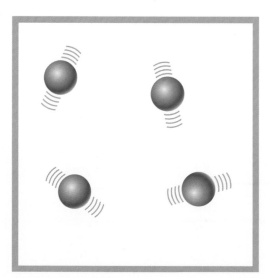

◄This diagram shows the particles in a gas.

What Happens When a Material is Heated?

This diagram shows what happens to the particles of a solid material as it is heated.

See how the particles move as the temperature rises

1

The particles in a solid vibrate slightly all the time.

2

As the solid warms up, the particles vibrate more. They take up more space. The solid expands.

3

Some particles break away from the solid. The solid is melting.

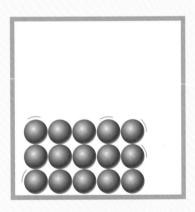

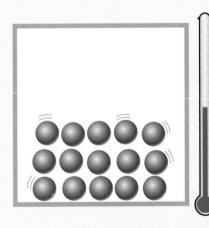

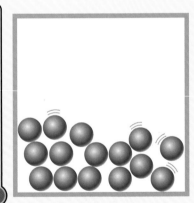

Melting Point

INCREASING

4

The material has now become a liquid.

5

The liquid is boiling. The particles are moving around faster. A few of the particles have enough energy to escape from the liquid. They evaporate.

6

When the temperature remains at boiling point for some time, all the liquid evaporates. The material has now become a gas.

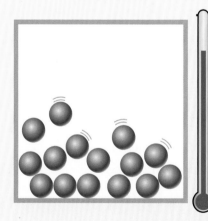

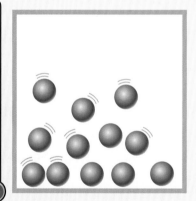

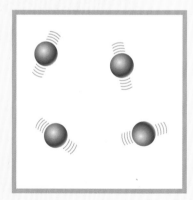

Boiling Point

TEMPERATURE

chapter eight

Ice: a Special Solid?

Why Does Ice Float?

Most materials get smaller as they cool down. We say they contract. This means that a lump of solid iron will not float in liquid iron.

Ice is different. Water contracts as you cool it down, but only until you get to 39°F (4°C). As it gets cooler than this, it starts to expand slightly. Ice forms at 32°F (0°C). Ice floats on water.

This property of ice is very important for animals and plants that live in water. If ice did not float, lakes, ponds and rivers would start to freeze from the bottom. Some might freeze completely. This could kill all the organisms living there.

But ice does float. A layer of ice forms on the top of a lake or pond. This ice stops the rest of the water from cooling too fast. Animals and plants can live in the liquid water beneath the ice.

▶(Opposite) Fish live in the water underneath the ice.

▲Icebergs floating on the sea.

Icebergs

The sea near the North Pole is covered in floating ice. The land at the South Pole is also covered in ice. Sometimes lumps of this ice break off and float away. Floating pieces of ice are called icebergs.

TITANIC DISASTER

Most of an iceberg is underwater. This makes it hard to tell the size of an iceberg by what we see. The passenger ship *Titanic* hit the underwater part of an iceberg in 1912. The
ice tore a hole in the ship, and it sank. About 1,500 people died.

Breaking Up Rocks

Water expands as it freezes. The rocks often have small cracks in them. Water runs into the cracks. When the temperature drops, the water freezes. As it freezes, it expands and pushes on the rock. This makes the crack wider. The ice melts when the weather warms up. Then it runs further down into the crack. This is repeated over and over again. Eventually, it can break up rocks.

▲These rocks were broken up by ice.

What Will Happen If the Ice Melts?

When we use energy or burn fuels, we add carbon dioxide to the atmosphere. This extra carbon dioxide makes Earth warmer.

Glaciers are "rivers" of ice that form in cold areas. Glaciers are getting smaller as the climate gets warmer. The ice at the North Pole and the South Pole is also melting. The water from melting ice at the South Pole is added to the sea. Sea levels around the world are rising.

▲ Will polar bears become extinct if the ice at the North Pole melts?

▲What will happen to the people on this island if the sea levels rise?

Someday, some low countries may be flooded. In many countries people who live near the coasts may have to move to higher land.

Glossary

aerosol (AIR uh sol) — mixture of droplets of liquid in a gas

atom (AT uhm) — tiny particles from which all materials are made

basalt (ba SAWLT) — type of rock formed when lava cools quickly

boiling (BOI ling) — when liquid is turning into a gas as fast as possible

boiling point (BOI ling point) — the temperature at which a liquid starts to boil

carbon dioxide (KAR buhn dye OK side) — a gas in the air that helps keep Earth warm

cloud (kloud) — small drops of water floating in the air

condensing (kuhn DENSS ing) — gas turning into liquid

crystal (KRISS tuhl) — a piece of material with sharp edges and flat surfaces

evaporating (i VAP uh rate ing) — liquid turning into gas

freezing (FREE zing) – liquid changing to a solid

freezing point (FREE zing point) — the temperature at which a liquid starts to freeze, the freezing point of a material is the same as its melting point.

gas (gass) — a substance that is invisible, easy to squash, and spreads out to fill the container that it is in

granite (GRAN it) — a type of rock that forms when magma cools slowly

humid (HYOO mid) — when there is a lot of water vapor in the air

lava (LAH vuh) — molten rock that comes out of a volcano

liquid (LIK wid) — a wet substance that you can pour

magma (MAG muh) — molten rock beneath Earth's surface

melting (MEL ting) — solid changing to a liquid

melting point (MEL ting point) — the temperature at which a solid starts to melt, the melting point of a material is the same temperature as its freezing point

microorganism (mye cro OR guh niz uhm) — a tiny living thing that can only be seen with a microscope

molecule (MOL uh kyool) — a tiny particle made from two or more atoms joined together

nitrogen (NYE truh juhn) — a gas in the air, about 78 percent of the atmosphere is nitrogen

oxygen (OK suh jun) — a gas in the air that we need to breathe, about 21 per cent of the atmosphere is oxygen.

pollution (puh LOO shuhn) — when harmful chemicals are added to the air, the ground, or the water

property (PROP ur tee) — a way of describing what something is like (for example, properties of ice are that it is hard and has a fixed shape)

solid (SOL id) — a substance that has a fixed shape and volume

state (state) — states of matter are solid, liquid, and gas

steam (steem) — water vapor formed when liquid boils

volume (VOL yuhm) — the amount of space that something takes up

water cycle (WAW tur SYE kuhl) — changes that happen to water as it evaporates from the sea, falls as rain, and then runs back into the sea

water vapor (WAW tur VAY pur) — water when it is a gas

Further information

Books

States of Matter. Carol Baldwin. Raintree, 2006.

States of Matter. Robert Snedden. Heinemann, 2007.

States of Matter: Gases, Liquids, and Solids. Krista West. Chelsea House, 2007.

Volcanoes. Michael Woods. Lerner, 2007.

Websites to visit

http://ga.water.usgs.gov/edu/mearth.html
USGS Water Science for schools.
This site offers information on many aspects of water, along with pictures, data, maps, and interactive activities.

http://www.kidzone.ws/water/
Information and activities on the water cycle.

http://www.weatherwizkids.com/lightning1.htm
This site provides some interesting facts about lightning.

http://saltthesandbox.org/rocks/index.htm
This site gives tips on identifying rocks in your neighborhood.

http://www.harcourtschool.com/activity/states_of_matter/
This shows an animation illustrating how particles are arranged in solids, liquids, and gases.

http://www.uscg.mil/lantarea/iip/students/default.htm
More information on icebergs, including a description of the journey of an iceberg.

http://www.epa.gov/kids/
U.S. Environmental Protection Agency.
This site provides more information on climate change. It also contains an animation showing the water cycle.

Index